UNDERGROUND STORAGE VAULTS

Protecting Priceless Information

by Kaitlyn Duling

CAPSTONE PRESS
a capstone imprint

Bright Ideas is published by Capstone Press, an imprint of Capstone.
1710 Roe Crest Drive
North Mankato, Minnesota 56003
www.capstonepub.com

Copyright © 2020 by Capstone. All rights reserved. No part of this publication may be reproduced in whole or in part, or stored in a retrieval system, or transmitted in any form or by any means, electronic, mechanical, photocopying, recording, or otherwise, without written permission of the publisher.

Library of Congress Cataloging-in-Publication Data
Names: Duling, Kaitlyn, author.
Title: Underground storage vaults : protecting priceless information / by Kaitlyn Duling.
Description: North Mankato, Minnesota : Capstone Press, 2020. | Series: High security | Includes bibliographical references and index. | Audience: Grades 4-6.
Identifiers: LCCN 2019029520 (print) | LCCN 2019029521 (ebook) | ISBN 9781543590586 (hardcover) | ISBN 9781543590593 (ebook)
Subjects: LCSH: Vaults (Strong rooms)—Juvenile literature. | Data protection—Juvenile literature.
Classification: LCC TH9734 .D85 2020 (print) | LCC TH9734 (ebook) | DDC 690/.143—dc23
LC record available at https://lccn.loc.gov/2019029520
LC ebook record available at https://lccn.loc.gov/2019029521

Image Credits
Alamy: CPC Collection, 21, Howard Barlow, 26–27, PA Images, 12–13, Rieger Betrand/Hemis, 11, 14, 28; AP Images: Dave Walsh/VWPics, cover, 17; Getty Images: Stephanie Strasburg/Bloomberg, 5, 6–7, 8–9, 18–19, 23; Shutterstock Images: Andrey_Popov, 24–25, Gelpi, 30–31
Design Elements: Shutterstock Images

Editorial Credits
Editor: Charly Haley; Designer: Laura Graphenteen; Production Specialist: Dan Peluso

All internet sites appearing in back matter were available and accurate when this book was sent to press.

Printed in the United States of America.
PA99

TABLE OF CONTENTS

CHAPTER 1

INSIDE IRON Mountain

The air is cool. The rooms are bright. Lockers line the hallways. Computers hum quietly. Guards keep watch for 24 hours a day. This is an underground storage **vault** in Boyers, Pennsylvania.

Lockers line the walls of a hallway at an underground storage vault.

Security guards help protect underground storage vaults.

The storage space is in an old **mine**. It protects items. It also stores information called **data**. Everything is 220 feet (67 meters) underground.

The vault is owned by a storage company called Iron Mountain. The company keeps many things safe.

Iron Mountain has many storage spaces. It stores things for people and businesses. The U.S. government uses Iron Mountain too.

Having a storage vault underground helps keep the things inside safe.

CHAPTER 2

UNDERGROUND Storage Vaults

The Iron Mountain mine is not the only underground storage space. There are others around the world. They all keep things **secure**. These things are protected.

SEED STORAGE

The Seed Vault is in Norway. It has more than 1 million seeds. They are stored deep in a mountain. They are kept frozen.

The Seed Vault in Norway holds seeds from many countries.

WHAT IS STORED?

People want to protect important items. Some of the items are paintings. Others are old photos or videotapes. Some have secret information. Underground storage spaces protect these things and more.

OLD PHOTOS

The Mormon Church has an underground storage space. Granite Mountain stores more than 3.5 billion photos.

Many old papers are stored in the Iron Mountain vault in Pennsylvania.

Workers bring items into the Seed Vault in Norway.

WORKERS

Thousands of people work in these storage areas. Some move items. Some are guards. Others work with computers. They protect data.

CHAPTER 3

LOCATION

Location is one way underground vaults keep items safe. These vaults are hard to reach. Some are in mines. These are hundreds of feet underground. Some are hidden in mountains.

The Seed Vault
is in a mountain
called Platåberget.

These locations keep items safe from weather. Rain and snow cannot get inside.

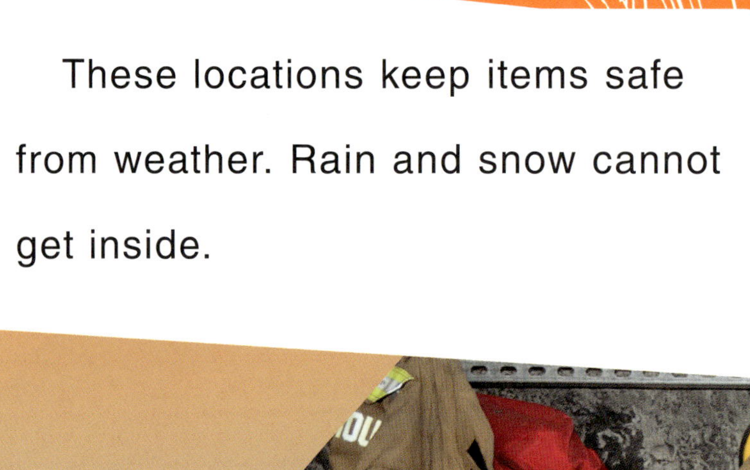

Underground storage vaults are hard to start on fire. But the Iron Mountain vault in Pennsylvania has its own fire department just in case.

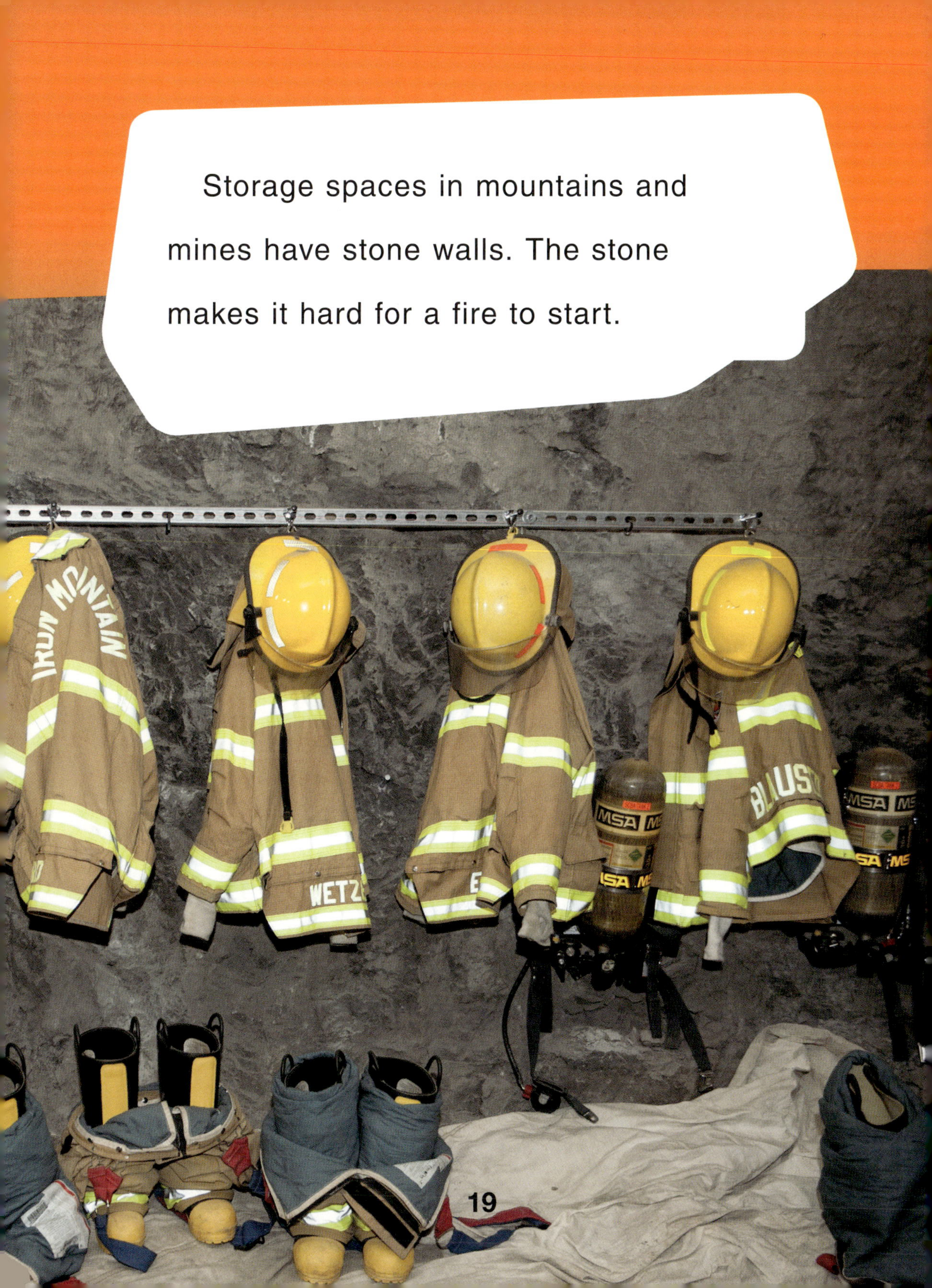

Storage spaces in mountains and mines have stone walls. The stone makes it hard for a fire to start.

CHAPTER 4

TEMPERATURE Control

Some items can become damaged if they get too hot. Others should not get too cold.

Underground storage spaces have controlled temperatures. This protects the items.

Heating and cooling systems keep underground storage vaults at the right temperature.

CHAPTER 5

LOCKS, Passwords, and Guards

Underground storage spaces have many ways to protect the items inside. They have guards and cameras. They have locked doors. Some have secret **passwords**.

THE CLOUD

Some storage spaces use the **cloud**. This is a way of storing data online. It is used for extra protection. The cloud is protected with passwords.

Big computers help Iron Mountain store data in the cloud.

GUARDS

Guards check each person who enters a vault. They make sure no one enters who is not supposed to. They make sure no one steals anything. The guards have weapons.

FINGERPRINT CHECK

Each person has different fingerprints. Guards at some storage spaces check fingerprints before people enter.

Security guards help protect underground storage vaults.

Locked doors are one of the many ways to keep important items safe underground.

Cameras record what happens in underground storage spaces. The videos can be used if something is stolen. They can help show who did it. But underground items are not stolen often.

Underground storage vaults keep things safe. They are protected.

GLOSSARY

cloud
a storage network that is connected over the internet

data
facts, numbers, and other information

mine
a pit in the ground from which coal, iron ore, and other natural materials are gathered

password
a secret word or code that lets people get into something

vault
a strongly built room or container with a lock used to store and protect valuable things

TRIVIA

1. One vault uses an underground lake to help keep its storage space cool. The lake covers 150 acres (60 hectares).

2. The temperature in the Iron Mountain mine is usually about 55 degrees Fahrenheit (12.7 degrees Celsius).

3. Vault walls in mines can be up to 30 feet (9 m) thick.

4. People made the Seed Vault by carving into solid rock.

ACTIVITY

KEEPING SOMETHING SAFE

Imagine something that might be stored in a guarded vault. It could be a very expensive item, important data, or anything else you need to keep safe. Write a story about this item. Why does this thing need to be stored in such a safe place? What secrets or value does it hold? Where do you think it would be stored in a facility? Will it be in a vault or on a computer? Will it be underground or somewhere else? Does it need to be stored at a certain temperature? Use examples from the book for ideas to help you create your story.

FURTHER RESOURCES

Interested in learning more about underground storage? Check out these websites:

Svalbard Global Seed Vault:
https://www.croptrust.org/our-work/svalbard-global-seed-vault

Wonderopolis: Where Is Cyberspace? What Is Cloud Computing?
https://www.wonderopolis.org/wonder/where-is-cyberspace

Want to know about more places that need high security? Check out these books:

Daly, Ruth. *Transportation Security Administration*. New York: Smartbook Media, Inc., 2019.

Harkrader, Lisa. *Fort Knox*. North Mankato, Minn.: Capstone Press, 2020.

Terrell, Brandon. *Guarding Area 51*. Mankato, Minn.: Child's World, 2016.

INDEX